FRIENDS OF TIME

时间的朋友

（汉英对照）

张恩祺　著

西北工業大學出版社
西　安

图书在版编目（CIP）数据

时间的朋友：汉英对照 / 张恩祺著 .-- 西安：西北工业大学出版社，2025.4.--ISBN 978-7-5612-9905-0

Ⅰ.P19-49

中国国家版本馆 CIP 数据核字第 2025MF9133 号

SHIJIAN DE PENGYOU

时 间 的 朋 友

张恩祺 著

责任编辑：李文乾　　**策划编辑**：华一瑾

责任校对：马婷婷　　**装帧设计**：高永斌　陈佳浩

出版发行：西北工业大学出版社

通信地址：西安市友谊西路 127 号　　邮　　编：710072

电　　话：（029）88491757　88493844

网　　址：www.nwpup.com

印 刷 者：西安五星印刷有限公司

开　　本：889mm×1194mm

印　　张：3.25

字　　数：47 千字

版　　次：2025 年 4 月第 1 版　　2025 年 4 月第 1 次印刷

书　　号：ISBN 978-7-5612-9905-0

定　　价：68.00 元

序言 Foreword

《时间的朋友》是一本独具匠心的小学生科普绘本，出自年轻的科学爱好者张恩祺同学之手。怀着对科普教育的热忱，她精心创作了该绘本，致力于把抽象的时间概念以生动具象的语言和图画直观地呈现出来。

我与张恩祺在一次科普活动中结识。作为时间科学领域的从业者，我时常接触到对科学满怀好奇的年轻人，但张恩祺对科学的那份热爱与对知识的渴求，依然令我印象深刻。她担任志愿者时展现出的专注和投入，让我看到了年轻一代对科学探索的执着。在与她交流时间测量、天文历法等专业知识的过程中，我发现她虽年纪尚小，却对时间的概念有着超越年龄的思考。

本书的创作灵感，源自张恩祺在中国科学院国家授时中心时间科学馆担任志愿者的经历。在带领小学生参观、讲解时间测量与认知历史的过程中，她深切体会到知识传递的重要性，遂决定以绘本形式普及时间科学。本书巧妙地将复杂的原理简单化，并搭配通俗易懂的插画，助力小读者揭开时间的神秘面纱。

本书不仅是张恩祺科学探索之旅的珍贵收获，也是一座连接儿童与科学知识的桥梁，让时间的魅力清晰可感，为新一代少年儿童播下热爱科学的种子。

张首刚

张首刚（1966—），男，研究员，中国科学院西安分院、陕西省科学院院长，中国科学院国家授时中心主任及首席科学家，长期从事时间频率研究。

目录 Contents

什么是时间? …… 2
What is time?

谁在大自然中记录时间? …… 8
Who records time in nature?

人类怎么记录时间? …… 12
How do humans record time?

傅科摆为什么能让人看到地球自转? …… 24
Why can the Foucault pendulum let people see the Earth's rotation?

古代计时仪器都有什么? …… 26
What are the ancient time-keeping instruments?

现代计时仪器有多厉害? …… 32
How powerful are modern time-keeping instruments?

授时技术经历了怎样的变化? …… 36
How has the time-transfer technology changed?

天文台是用来干什么的? …… 40
What is an observatory for?

真的可以时空旅行吗? …… 44
Is time travel really possible?

"What time is it now?" or "How long I've been waiting for you".

我们如何知道时间呢？其实，我们是通过观察一些自然现象，比如地球自己转一圈需要多久，月亮绕着地球转一圈需要多久，地球绕着太阳转一圈需要多久，这些周期性的过程来确定时间的。

How do we know time? In fact, we determine time by observing some natural phenomena, such as how long it takes for the Earth to rotate on its own, how long it takes for the Moon to orbit the Earth, and how long it takes for the Earth to orbit the Sun. These regular changes help us determine time.

简单来说，时间是通过观察大自然中的规律来计算的。

In short, time is calculated by observing the laws of nature.

什么是时间？

What is time?

很多科学家认为，大约在 138 亿年前，宇宙从一场大爆炸中诞生了。从此以后，就有了时间。

Many scientists believe that about 13.8 billion years ago, the universe was born from a big bang. Since then, time has existed.

什么是时间？

What is time?

时间是物理学中的七个基本物理量之一，符号为 t。时间的基本单位是秒，符号为 s。

Time is one of the seven fundamental physical quantities in physics, with the symbol t. The basic unit of time is the second, with the symbol s.

生活中常用的时间单位还有毫秒、分钟、小时、日、月、年等。

Commonly used time units in life also include milliseconds, minutes, hours, days, months, years, etc.

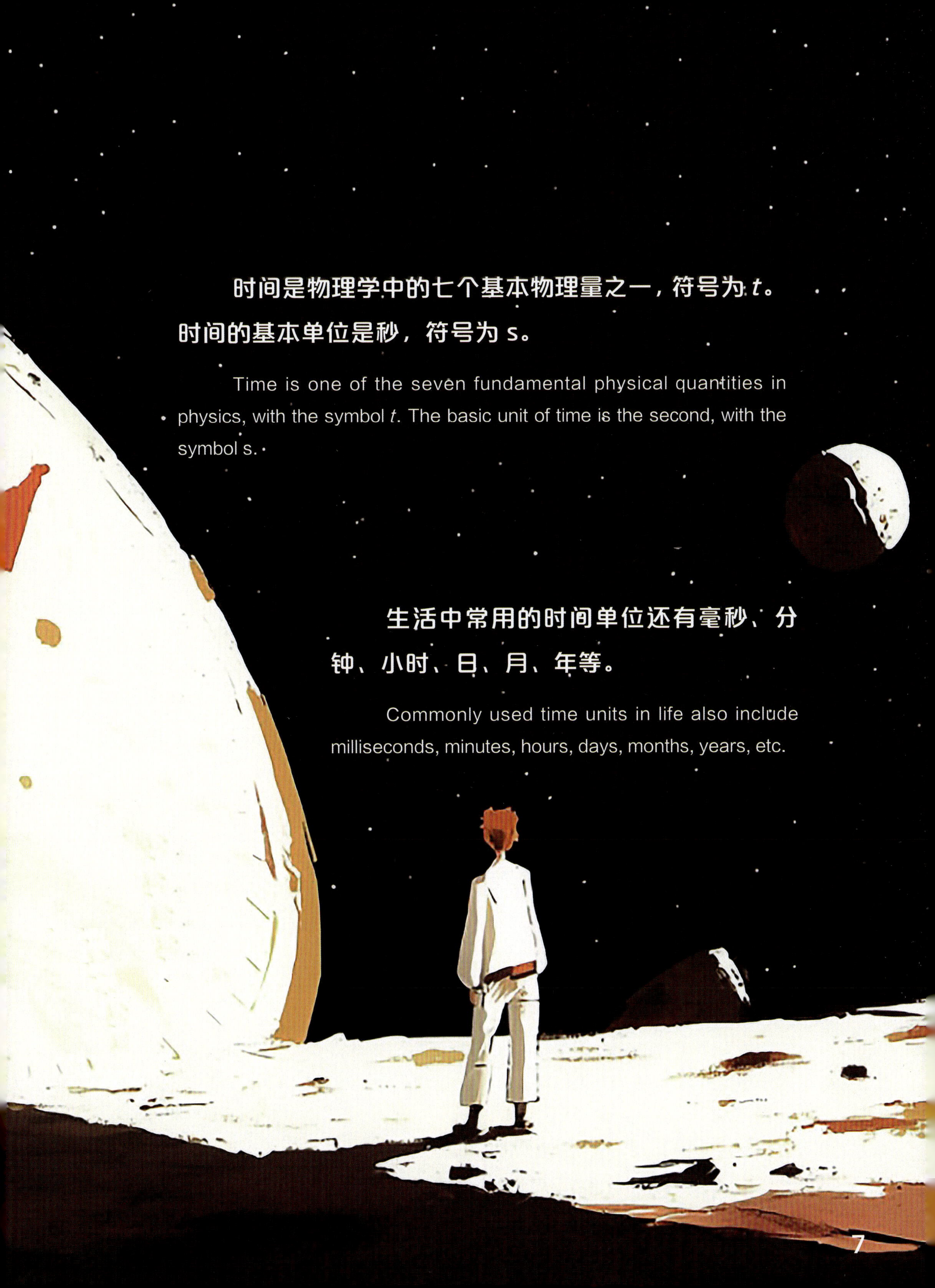

化 石
Fossils

地球上的化石形成于不同时期，它们有的已经存在几亿年了。根据化石的成分，人们可以分析出当时的自然环境和生物信息。

其中最著名的是澄江生物群，它位于我国云南澄江帽天山附近，是保存完整的寒武纪早期古生物化石群。它生动地再现了 5.3 亿年前海洋生命的壮丽景观和现生动物的原始特征，为研究地球早期生命的起源、演化、生态适应等理论提供了珍贵证据。

Fossils on the Earth were formed in different periods, and some of them have existed for hundreds of millions of years. Based on the composition of fossils, people can analyze the natural environment and biological information at that time.

The most famous among them is the Chengjiang Biota. It is located near Maotianshan Mountain in Chengjiang, Yunnan Province, China, and is a well-preserved paleontological fossil group from the early Cambrian period. It vividly reproduces the magnificent landscape of marine life 530 million years ago and the primitive characteristics of extant animals, providing precious evidence for the study of theories such as the origin, evolution, and ecological adaptation of life in the early days of the Earth.

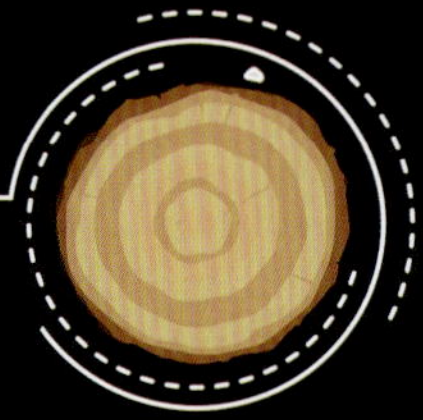

年 轮
Annual Rings

一个年轮代表着树木经历了所生长环境的一个周期的变化。通常气候是一年一个变化周期，所以年轮也就代表着树木一年中的生长情况。根据年轮的数目，可以推知树木的年龄，考察森林的年代。

An annual ring represents the tree's experience of a cycle of changes in its growing environment. Usually, the climate undergoes a cycle of change every year. Therefore, the annual ring also represents the growth pattern of the tree in a year. Based on the number of annual rings, the age of the tree can be deduced, and the age of the forest can be estimated.

红珊瑚
Red Coral

红珊瑚属有机宝石，色泽红润，质地坚韧，由珊瑚虫堆积而成，生长极为缓慢，大约 20 年才生长 2.54 厘米，300 年左右才能长成大约 1 千克，因而有“千年珊瑚万年红”之说。

Red coral is an organic gemstone. It has a ruddy and glossy color and a tough texture. It is formed by the accumulation of coral polyps. Its growth is extremely slow. It takes approximately 20 years for it to grow 2.54 centimeters, and it takes about 300 years to grow into a piece weighing about 1 kilogram. Therefore, there is a saying that "coral takes a thousand years to form and turns red after ten thousand years".

人类怎么记录时间?

How do humans record time?

我们习惯用"年、月、日、时、分、秒"来记录时间。

We are accustomed to using "year, month, day, hour, minute and second" to record time.

如果把地球存在的漫长时间看作一天，那么直到现在，人类出现的时间都还不到两分钟。记录非常久远的时间，需要采用特殊的方式——大尺度时间。

If we regard the long period of time during which the Earth has existed as one day, then up to now, the time since the emergence of human beings is less than two minutes.To record extremely long periods of time, we need to adopt a special method — large-scale time.

大尺度时间
Large-scale Time

根据"宇宙大爆炸"理论，宇宙起源于大约 138 亿年前的一次大爆炸。这是至今人类能了解到的最大时间尺度。

According to the "Big Bang" theory, the universe originated from a big bang about 13.8 billion years ago. This is the largest time scale that humans can understand so far.

如果把 138 亿年的宇宙演化史浓缩为 1 年，那么 1 天≈ 3800 万年，1 秒≈ 440 年，人的寿命≈ 0.2 秒。在宇宙年代中，我们"转瞬即逝"。

If we condense the 13.8-billion-year cosmic evolution history into a year, one day ≈ 38 million years, one second ≈ 440 years, and a human lifespan ≈ 0.2 seconds. In the cosmic time scale, we "disappear in a flash".

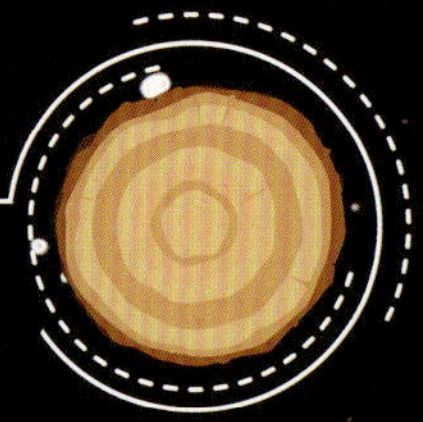

宇宙年历
Cosmic Calendar

- **大约 138 亿年前，宇宙大爆炸。**
  About 13.8 billion years ago, the Big Bang occurred.
- **大约 136 亿年前，第一批恒星形成。**
  About 13.6 billion years ago, the first stars formed.
- **大约 110 亿年前，银河系形成。**
  About 11 billion years ago, the Milky Way Galaxy formed.
- **大约 46 亿年前，太阳系形成。**
  About 4.6 billion years ago, the Solar System formed.
- **大约 25 亿年前，地球上有氧大气出现。**
  About 2.5 billion years ago, an oxygen-rich atmosphere appeared on Earth.
- **大约 20 亿年前，地球生物有性繁殖出现。**
  About 2 billion years ago, sexual reproduction of organisms on Earth emerged.
- **大约 5.4 亿年前，寒武纪生物大爆发。**
  About 540 million years ago, the Cambrian Explosion occurred.
- **大约 4.4 亿年前，陆地蕨类植物出现。**
  About 440 million years ago, terrestrial ferns appeared.

- **大约 4.4 亿年前，裸子植物、爬行动物出现。**
  About 440 million years ago, gymnosperms and reptiles emerged.
- **大约 2.5 亿年前，恐龙出现。**
  About 250 million years ago, dinosaurs appeared.
- **大约 6500 万年前，恐龙灭绝。**
  About 65 million years ago, dinosaurs went extinct.
- **大约 700 万年前，最早古人类出现。**
  About 7 million years ago, the earliest hominids emerged.

地质年代
Geological Time

地质年代是用来描述地球历史不同阶段的长时间尺度单位。

The geological time scale is a unit of long time scale used to describe different stages of the Earth's history.

依据岩石、岩层的形成与生物演化的顺序，地壳发展历史可划分为若干个自然阶段。

According to the sequence of the formation of rocks and rock strata and the evolution of organisms, the history of the development of the Earth's crust is divided into several natural stages.

例如：

For example:

- **古生代（距今约 5.4 亿～ 2.5 亿年）：海生无脊椎动物大量出现，陆地上有了裸蕨等植物。**

Paleozoic Era (about 540 million – 250 million years ago): Marine invertebrates appeared in large numbers, and plants like bare ferns emerged on land.

- **中生代（距今约 2.5 亿～ 6500 万年）：分为三叠纪、侏罗纪、白垩纪。**

Mesozoic Era (about 250 million – 65 million years ago): It is divided into the Triassic, Jurassic, and Cretaceous periods.

三叠纪：珊瑚、海龟等出现了。

Triassic period: Corals, sea turtles, etc. appeared.

侏罗纪：陆地上的裸子植物非常茂盛，恐龙是地球的霸主。

Jurassic period: Gymnosperms on land were very prosperous, and dinosaurs dominated the Earth.

白垩纪：在白垩纪末期，行星、彗星撞击地球，许多植物、动物都灭绝了。

Cretaceous period: At the end of the Gretaceacs period, planets and comets hit the Earth, leading to the extinction of many plants and animals.

- **新生代（6500 万年前至今）：哺乳动物大爆发，人类文明开始发展。**

Cenozoic Era (65 million years ago to the present): There was a mammalian explosion, and human civilization began to develop.

历史年代
Historical Eras

历史年代是描述和划分人类历史上具体时期的时间单位。

Historical eras are time units used to describe and divide specific periods in human history.

例如：

For example:

- **世纪：常用的历史时间单位，一个世纪包含 100 年。**
  Century: It is a commonly used historical time unit and a century contains 100 years.
- **年代：更精确地指一个世纪中的特定十年，如 20 世纪 80 年代。**
  Decade: It more precisely refers to a specific ten-year period within a century, such as the 1980s.
- **朝代：特别在中国历史中，朝代的更迭常常用来标记历史年代的变迁。**
  Dynasty: Especially in Chinese history, the succession of dynasties is often used to mark the changes of historical eras.

人类怎么记录时间？

How do humans record time?

天文历法

Astronomical Calendars

天文历法是一种根据天体的运动和相对位置来计算时间的方法，它涵盖了日历的制定和各种时间单位的划分。

Astronomical calendars are methods of calculating time based on the movements and relative positions of celestial bodies. They cover the formulation of calendars and the division of various time units.

阳历和阴历

Solar Calendar and Lunar Calendar

阳历又称太阳历，以地球绕太阳的公转周期为基础，主要协调年和日的关系，准确反映四季变化。

The solar calendar, also known as the sun-based calendar, is based on the Earth's revolution period around the Sun. It mainly coordinates the relationship between years and days and accurately reflects the changes of the four seasons.

阴历又称月亮历，基于月亮绕地球的周期，主要协调月和日的关系，准确反映月相盈亏变化。

The lunar calendar, also known as the moon-based calendar, is based on the Moon's cycle around the Earth. It mainly coordinates the relationship between months and days and accurately reflects the changes of the lunar phases.

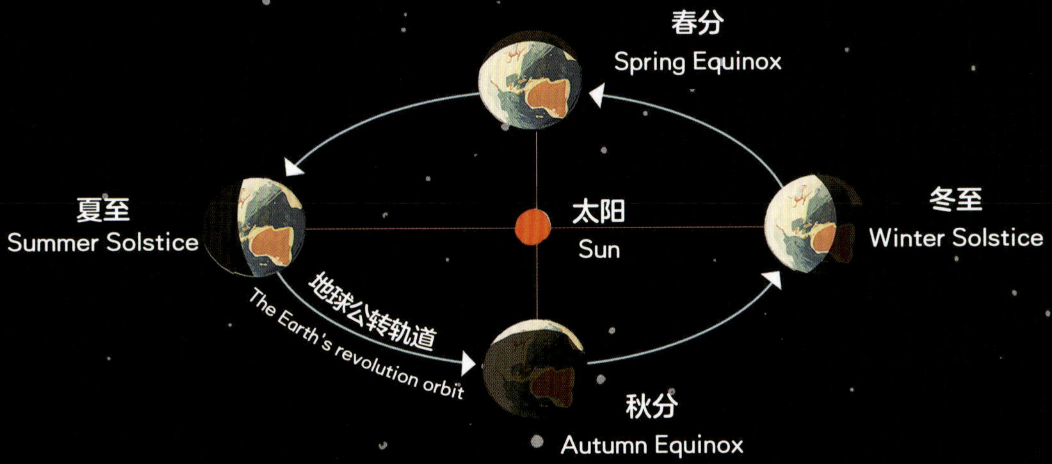

地球绕太阳公转形成四季

The Earth's revolution around the Sun forms the cycle of the four seasons

二十四节气
24 Solar Terms

在浩瀚的天文知识宝库中，二十四节气是中华民族古老智慧的结晶。它宛如一条精妙的丝线，巧妙地串联起天文、气象与农事活动。

In the vast treasure trove of astronomical knowledge, the 24 Solar Terms are a remarkable example of the ancient wisdom of the Chinese nation. They are like an exquisite thread that weaves together astronomy, meteorology, and agricultural activities.

二十四节气的形成，与地球在公转轨道上的位置密切相关。地球绕太阳公转一周，约为 365.24 天，古人将这个公转轨道划分为 24 等份，每一份对应一个节气，相邻两个节气间隔约 15 天。太阳直射点在地球不同纬度的移动，带来了气候、物候的显著变化，二十四节气便依此有序更迭。

The formation of the 24 Solar Terms is closely related to the position of the Earth in its orbit around the Sun. The Earth takes approximately 365.24 days to complete one orbit around the Sun. The ancients divided this orbit into 24 equal parts, with each part corresponding to a solar term, and the interval between adjacent solar terms is about 15 days. As the solar declination moves across different latitudes of the Earth, it brings significant changes in climate and phenology, and the 24 Solar Terms change in an orderly manner accordingly.

二十四节气可分为四类：表示寒来暑往变化的有立春、春分、立夏、夏至、立秋、秋分、立冬、冬至八个节气。象征气温变化的有小暑、大暑、处暑、小寒、大寒五个节气。反映降水量的有雨水、谷雨、白露、寒露、霜降、小雪、大雪七个节气。反映物候现象或农事活动的有惊蛰、清明、小满、芒种四个节气。

The 24 Solar Terms can be categorized into four groups. The eight terms that indicate the changes of seasons are: Beginning of Spring, Spring Equinox, Beginning of Summer, Summer Solstice, Beginning of Autumn, Autumn Equinox, Beginning of Winter, and Winter Solstice. The five terms that symbolize temperature changes are: Lesser Heat, Greater Heat, End of Heat, Lesser Cold, and Greater Cold. The seven terms that reflect precipitation are: Rain Water, Grain Rain, White Dew, Cold Dew, First Frost Descent, Light Snow, and Heavy Snow. The four terms that reflect phenological phenomena or agricultural activities are: Waking of Insects, Pure Brightness, Grain Buds, and Grain in Ear.

傅科摆为什么能让人看到地球自转?

Why can the Foucault pendulum let people see the Earth's rotation?

为了证明地球在自转，1851 年，法国物理学家傅科做了一次成功的摆动实验，傅科摆由此而得名。

In 1851, in order to prove the Earth's rotation, French physicist Léon Foucault conducted a successful pendulum experiment, from which the Foucault pendulum got its name.

地球自转
Earth's rotation

上海天文馆的傅科摆
The Foucault pendulum at the Shanghai Astronomy Museum

实验在法国巴黎先贤祠最高的圆顶下方进行，摆长 67 米，摆锤重 28 千克，悬挂点经过特殊设计使摩擦降到最低。

The experiment was carried out under the highest dome of the Panthéon in Paris, France. The pendulum was 67 meters long, the pendulum bob weighed 28 kilograms, and the suspension point was specially designed to minimize

在傅科摆试验中，人们看到单摆每次摆动的方向都会稍稍偏离原轨迹。这说明摆平面和地球发生了相对运动，从而有力地证明了地球在自转。

In the Foucault pendulum experiment, people observed that the direction of each swing of the pendulum deviated slightly from the original trajectory. This indicates that there is relative motion between the plane of the pendulum and the Earth, thus powerfully proving that the Earth is rotating.

古代计时仪器都有什么？

What are the ancient time-keeping instruments?

太阳钟

Sun Clock

太阳钟有很多种，我们最熟悉的有圭表、英国巨石阵、日晷等。它们都是通过观察太阳投影的变化来记录时间的。

There are many types of sun clocks. The most familiar ones to us are the gnomon, Stonehenge in the UK, the sundial, etc. They all record time by observing the changes of the sun's shadow.

圭表

Gnomon

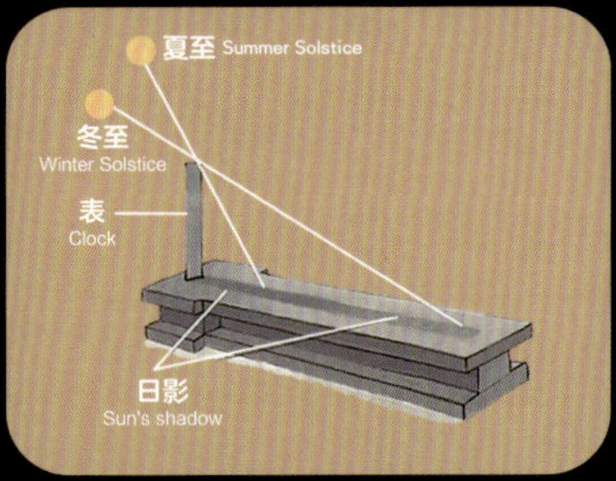

圭表是一种古老的测量日影长度的天文仪器，由直立的竿（表）和一个南北方向水平放置的尺（圭）组成，用以观察太阳光投射的竿影影长和变化规律，以定午时和四时节气等。

The gnomon is an ancient astronomical instrument for measuring the length of the sun's shadow. It consists of an upright pole (the "biao") and a horizontally placed ruler (the "gui") in the north-south direction. It is used to observe the length and change rules of the shadow of the pole projected by sunlight, so as to determine noon and the solar terms of the four seasons.

英国巨石阵

Stonehenge in the UK

每年夏至，太阳都会从特定的标石上升起，因此很多人都相信，巨石阵其实是当时的人们用来测算历法的工具。人们通过观测日出、日落等天象以定节气时令，以此指导农耕。

Every year on the Summer Solstice, the sun rises from a specific marker stone. Therefore, many people believe that Stonehenge was actually a tool used by people at that time to calculate the calendar. They observed celestial phenomena such as the sunrise and sunset to determine the solar terms and guide farming.

日晷

Sundial

中国是世界上最早使用日晷进行计时的国家之一。日晷由晷针和晷面组成。当太阳光照射日晷时，晷面上就会出现铜针的影子。人们观察针影在晷面上的位置，这样就能知道不同的时间啦。

China is one of the earliest countries in the world to use the sundial for time-keeping. It is composed of a bronze needle and a dial face. When the sunlight shines on the sundial, the shadow of the bronze needle will appear on the dial face. People observe the position of the needle's shadow on the dial face, and in this way, they can tell different times.

古代计时仪器都有什么？

What are the ancient time-keeping instruments?

火　钟

Fire Clock

人们发现，一定数量的同一种燃料，比如蜡烛、灯油、香……其燃烧速度基本不变，因此均匀的燃烧过程可以用于计时，于是发明了火钟。火钟的计时精度不高，但很方便。

People discovered that for a certain amount of the same kind of fuel, for example, candles, lamp oil, incense, ect., the burning speed is basically constant, and the uniform burning process can be used for time-keeping. Thus, the fire clock was invented. Although the fire clock is not highly accurate in time-keeping, it is very convenient.

蜡钟
Wax clock

油灯钟
Oil-lamp clock

香钟
Incense clock

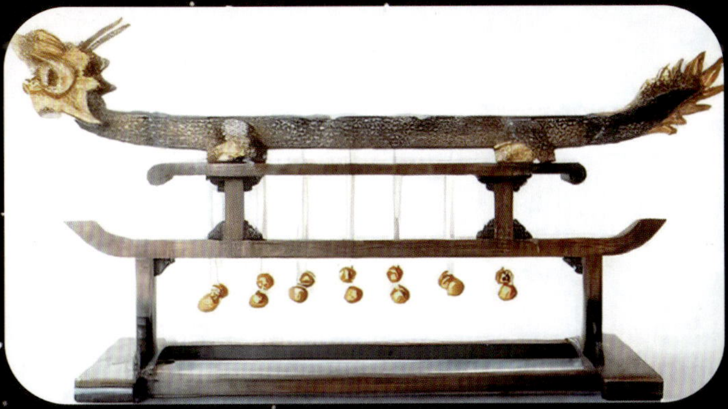

龙舟香漏
Dragon-Boat-shaped
Incense-leak clock

龙舟香漏的主体是一艘龙舟形的盛器，通常由金属制成。龙舟上放置一至两根燃着的香，香上横着数条两端系有金属球的线。香在燃烧过程中，会逐渐烧断横在上面的线。当线被烧断时，金属球就会跌进下面的盛器，发出响声，以此来报告时间。

The main structure is a boat-shaped container in the form of a dragon boat, usually made of metal. On the dragon boat, there is one or two incense sticks that are lit. Several lines with metal balls tied at both ends are stretched across the incense sticks. As the incense burns, it gradually severs these lines. When a line is burned through, the metal ball falls into the container below, making a noise to signal the passage of time.

机械钟
Mechanical Clock

1582—1583 年，意大利物理学家和天文学家伽利略发现了摆的等时性。摆钟是一种利用摆锤的周期性振动来计量时间的机械钟。

From 1582 to 1583, the Italian physicist and astronomer Galileo discovered the isochronism of the pendulum. A pendulum clock is a kind of mechanical clock that measures time by utilizing the periodic vibration process of the pendulum bob.

时间 = 摆的振动周期 × 振动次数

Time = Vibration period of the pendulum × Number of vibrations

以摆作为振动系统的钟通常都带有报时功能，所以又称自鸣钟。

Clocks with a pendulum as the vibration system usually have a time-reporting function, so they are also called chiming clocks.

1657 年，荷兰物理学家和天文学家克里斯蒂安 · 惠更斯利用摆的等时性原理发明了摆钟。后不断改进，沿用至今。

In 1657, the Dutch physicist and astronomer Christiaan Huygens invented the pendulum clock by using the principle of the isochronism of the pendulum. After continuous improvements, it has been used until now.

现代计时仪器有多厉害？

How powerful are modern time-keeping instruments?

电子钟、电波钟、原子钟、脉冲星计时……这些计时工具的发明，让计时精准度大大提高。

Electronic clock, radio-controlled clock, atomic clock, pulsar time-keeping... The invention of these time-keeping tools has greatly improved the accuracy of time-keeping.

电子钟

Electronic Clock

电子钟是一种利用数字电路来显示秒、分、时的计时装置，与传统的机械钟相比，它具有走时准确、显示直观、无机械传动装置等优点，因而得到广泛应用。智能手表就是具有无线信息接收和处理能力并具有手表基本功能的智能穿戴设备。

An electronic clock is a time-keeping device that uses digital circuits to display seconds, minutes, and hours. Compared with traditional mechanical clocks, it has the advantages of accurate time-keeping, intuitive display, and no mechanical transmission device, so it is widely used. Smartwatches are intelligent wearable devices with the ability to receive and process wireless information and the basic functions of a watch.

电波钟
Radio-controlled Clock

电波钟由电子时钟和无线电接收系统组成，通过无线电接收系统接收由国家授时中心发出的准确时间，经 CPU 处理后将时间显示出来。电波钟的精度是万分之一秒。

A radio-controlled clock consists of an electronic clock and a radio receiving system. The accurate time is issued by the national time-service center. It is received by the radio receiving system, processed by the CPU, and then the time is displayed. The accuracy of a radio-controlled clock is one-ten thousandth of a second.

原子钟

Atomic Clock

原子钟是一种计时装置，它最初是由物理学家创造出来用于探索宇宙本质的。他们从来没有想过这项技术有朝一日竟能应用于全球导航系统上。

原子钟 2000 万年才误差 1 秒。

An atomic clock is a time-keeping device. It was originally created by physicists to explore the essence of the universe. They never thought that this technology would one day be used in global navigation systems.

An atomic clock has an error of only one second in 20 million years.

脉冲星计时

Pulsar Timing

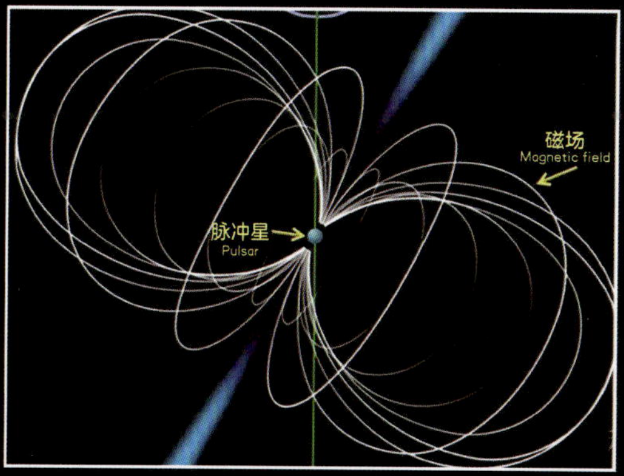

脉冲星计时是以毫秒脉冲星的自转周期为基准建立的时间计量系统，具有超稳定的时间频率特征，是极端物理条件太空实验室和宇宙中最稳定、最精确的天文钟。

Pulsar timing is a time measurement system established based on the rotation period of millisecond pulsars. It has the characteristic of an ultra-stable time frequency and serves as the most stable and accurate astronomical clock in the space laboratory under extreme physical conditions and in the universe.

授时技术经历了怎样的变化？

How has the time-transfer technology changed?

确定、保持某种时间尺度，通过一定的通信手段把这种时间尺度的信息传送出去，供应用者使用，这一整套工作称为授时，也称作时间服务。

从古至今，授时技术发生着巨大的变化。

Determining and maintaining a certain time scale, and transmitting the information of this time scale through certain communication means for users to use. This whole set of work is called time-transfer, also known as time-service.

From ancient times to the present, the time-transfer technology has undergone huge changes.

鸡人呼旦

The "Chicken Man" announcing dawn

中国古代皇宫内侍卫头戴鸡冠状赤色头巾，身穿绿色衣袍，于清晨报晓。

In ancient China, the guards in the imperial palace wore red cockscomb-shaped headscarves and green robes, announcing the dawn in the early morning.

打更报时

Beating the night watch to announce the time

中国古代的夜间报时制度，每更一个时辰，对应现在两个小时。

It was a night-time-announcing system in ancient China. Each watch corresponded to one shichen, which is equivalent to two hours nowadays.

钟鼓报时

Announcing time by bells and drums

中国古代曾经把敲钟和击鼓作为一个城市的报时手段。

In ancient China, ringing bells and beating drums were once used as a means of announcing the time in cities.

塔钟报时

Tower clock time-announcing

塔钟是安装在塔状建筑物上的大型钟，向公众报时。

A tower clock is a large-scale clock installed on a tower-shaped building, announcing the time to the public.

无线电授时

Radio time transfer

无线电授时，覆盖范围广，授时精度高，有长波授时、短波授时、低频时码授时。

Radio time transfer has a wide coverage and high time-transfer accuracy. There are long-wave time transfer, short-wave time transfer, and low-frequency time-code time transfer.

卫星授时

Satellite time transfer

全球已形成美国全球定位系统、俄罗斯格洛纳斯、中国北斗和欧盟伽利略四大卫星导航系统。卫星导航系统精度高、全天时、全天候、大范围、成本低，是当今标准时间频率信号传递的主要手段之一。

Globally, four major satellite navigation systems have been formed: GPS in the United States, GLONASS in Russia, Beidou in China, and Galileo in the European Union. Satellite navigation systems feature high precision, all-day and all-weather operation, wide coverage, and low cost. They are one of the main means of transmitting standard time-frequency signals today.

互联网授时

Internet time transfer

利用互联网传递标准时间，为网络计算机和各种网络终端提供时间同步信息。

It uses the Internet to transmit standard time and provides time-synchronization information for network computers and various network terminals.

落球报时

Time-announcing by dropping balls

落球报时是在重要的码头、港口竖起高杆，在顶端挂上圆球，按规定时升落圆球，借以向沿海人员报告时间。

For time-announcing by dropping balls, tall poles were erected at important docks and ports. A ball was hung at the top, and it was raised and lowered at specified times to inform people along the coast of the time.

天文台是用来干什么的?

What is an observatory for?

时间在巨型射电望远镜的波纹中流淌，于原子钟的震颤里结晶，而天文台始终是丈量永恒的最初与最终标尺。

Time flows in the ripples of the giant radio telescope and crystallizes in the tremors of the atomic clock. And the observatory has always been the first and final yardstick for measuring eternity.

天文台是专门进行天象观测和天文学研究的机构。世界各国天文台大多设在山上。每个天文台都拥有一些观测天象的仪器设备，主要是天文望远镜。

An observatory is an institution specifically dedicated to astronomical observation and astronomical research. Most observatories around the world are located on mountains. Each observatory is equipped with some instruments and equipment for observing celestial phenomena, mainly astronomical telescopes.

光学天文台
Optical Observatory

光学天文台主要装备各光学天文仪器，如光学天文望远镜、太阳镜等，用于方位天文学或天体物理学方面的研究。

It is mainly equipped with various optical astronomical instruments, such as optical astronomical telescopes and solar telescopes, and engages in research in positional astronomy or astrophysics.

天文台是用来干什么的?

What is an observatory for?

射电天文台

Radio Observatory

射电天文台一般由巨型甚至超巨型的无线接受设备和基站等构成，装备射电望远镜，观察的范围更大，受干扰小，用于射电天文学的研究。

Generally, it is mainly composed of giant or even super-giant wireless receiving devices and base stations. It is equipped with radio telescopes, has a wider observation range, is less affected by celestial interference, and conducts research in radio astronomy.

20 世纪 60 年代，天文学取得了四项非常重要的发现：脉冲星、类星体、宇宙微波背景辐射、星际有机分子。它们被称为 “四大发现”。这四项发现都与射电望远镜有关。

In the 1960s, four very important discoveries were made in astronomy: pulsars, quasars, cosmic microwave background radiation, and interstellar organic molecules, which are known as the "Four Great Discoveries". All four of these discoveries are related to radio telescopes.

空间天文台
Space Observatory

空间天文台由一些用于空间观测的人造卫星组成，配备非常先进的光学观测系统。

It consists of artificial satellites and is equipped with very advanced optical observation systems.

真的可以时空旅行吗？

Is time travel really possible?

在各种科幻小说中，时光旅行是最令人激动的设想之一。人们想象在登上时光机器后，可以在瞬间前往他所希望的那个时代。对于人类是否真的可以进行时空旅行，有很多大胆的假设……

In various science-fiction novels, time travel is one of the most exciting concepts. People imagine that after boarding a time-machine, they can instantly travel back to the era they desire. There are many bold hypotheses about whether humans can actually travel through time and space...

接近光速飞行

Flying at Near-light Speed

根据爱因斯坦相对论，物体运动的速度越快，时间流逝就越慢。一旦物体运动的速度达到光速，时间就等于零，也就意味着时间静止了！

According to Einstein's theory of relativity, the faster the speed of an object, the slower the time passes. Once an object moves at the speed of light, time stops, which means time stands still!

若你乘坐一艘飞行速度接近光速的飞船在太空中飞行一个星期，这一个星期相当于地球上的几年时间。飞行结束后你再回到地球，就相当于进入了未来。但以接近光速飞行，理论上可行，实际上是做不到的。

Imagine you boarded a spaceship that could fly at near-light speed in space for a week. The week in space would be equivalent to several years on Earth. If you return to Earth, you would be traveling into the future.However, although flying at near-light speed is theoretically possible, it is actually unattainable.

虫洞穿越
Wormhole Travel

人们设想，虫洞可以将两个不同的时空连接在一起。相连的两个时空，可以是两个完全不同的宇宙，也可以是同一宇宙的两个不同地方。任何物体都可以通过虫洞，从一个时空自由出入另一个时空。

People assume that wormholes can connect two different spacetimes. The two connected spacetimes can be two completely different universes or two different locations in the same universe. Any object can freely move through a wormhole from one spacetime to another.

想象一下，在公元 3000 年的人类意识上传中心，人类复制自己的意识，并将其输入一个空白的纳米机器人。这个纳米机器人乘坐飞船，穿过一个虫洞，就相当于自己到了另一个时空。

Imagine that in the year 3000, at a human consciousness-uploading center, people duplicate their minds and transfer them into blank nanobots. These nanobot carriers then board a spacecraft, traverse a wormhole, and effectively arrive in another spacetime—as if they had traveled there themselves.

虫洞连接着未来，也连接着过去。但我们也别太乐观了，因为为时间机器提供能量的装置绝不是人类的技术所能达到的。

Wormholes connect the future and the past. However, we shouldn't be too optimistic because the device that provides power to a time-machine is far beyond the reach of human technology.